21ST Century Skills INNOVATION LIBRARY

Exploring the Internet of Things

Smart Wearable Devices

Martin Gitlin

Cherry Lake Press

Published in the United States of America by Cherry Lake Publishing
Ann Arbor, Michigan
www.cherrylakepublishing.com

Reading Adviser: Marla Conn, MS, Ed., Literacy specialist, Read-Ability, Inc.

Photo Credits: © Rawpixel.com/Shutterstock, cover, 1; © Dan Rentea/Shutterstock, 5; © Dusan Petkovic/Shutterstock, 8; © Kathy Hutchins/Shutterstock, 11; © Ceri Breeze/Shutterstock, 13; © Onfokus/iStock, 14; © Yugong Luo/Shutterstock, 15; © Maurizio Pesce/WikiMedia, 16; © Odua Images/Shutterstock, 19; © garetsworkshop/Shutterstock, 21; © Gorynvd/Shutterstock, 22; © HQuality/Shutterstock, 23; © Grigorita Ko/Shutterstock, 24; © LightField Studios/Shutterstock, 27; © dvoevnore/Shutterstock, 29

Graphic Element Credits: © Ohn Mar/Shutterstock.com, back cover, multiple interior pages;
© Dmitrieva Katerina/Shutterstock.com, back cover, multiple interior pages;
© advent/Shutterstock.com, back cover, front cover, multiple interior pages;
© Visual Generation/Shutterstock.com, multiple interior pages;
© anfisa focusova/Shutterstock.com, front cover, multiple interior pages;
© Babich Alexander/Shutterstock.com, back cover, front cover, multiple interior pages

Library of Congress Cataloging-in-Publication Data has been filed and is available at catalog.loc.gov.

Printed in the United States of America
Corporate Graphics

Table of Contents

Introduction 4

Chapter one
Smart Wearable Devices 6

Chapter two
Drawbacks to Smart Wear 12

Chapter three
The Future of Wearable Devices 20

Critical Thinking and Problem Solving 28
Learn More 30
Glossary 31
Index 32

Introduction

Glasses were created to improve sight. Watches were invented to keep time. Clothes were designed to keep people comfortable, yet stylish. But the role of these wearables has never gone beyond basic uses—until now, with the introduction of the Internet of Things (IoT) **technology**. IoT has helped reinvent a number of industries and things, including wearables.

IoT technology connects everyday objects to the internet through the use of special **sensors**. The technology links physical objects to computer-based systems. And it is starting to impact the world of wearables from head to toe.

There are three layers of IoT technology at work in wearables. The first layer is the sensors. These sensors are placed directly on the body. They **monitor** things such as temperature, movement, and pulse. The second layer is the connection between the wearable and a smartphone or home network. The

In the United States, more than half of the people own a smart wearable device.

third layer is the **cloud**. This is where the wearable device sends, reads, and analyzes the **data** collected.

IoT wearables affect many aspects of life. They allow users to monitor and improve health. They make gaming more physically demanding and complex. They help teachers understand and motivate students. And this only scratches the surface. The future of IoT wearables seems filled with endless possibilities. Can this technology truly change lives in ways never imagined?

CHAPTER ONE

Smart Wearable Devices

IoT technology has a wide range of benefits. It aids the healthcare industry by its ability to detect and prevent disease. The technology helps farmers feed the world. IoT makes vehicles and roads safer for traveling. It even benefits those who are health conscious, fashionable, or into video gaming!

Some wearable devices promote good health. Among them are Fitbit and other activity trackers and smartwatches. They let users know how many steps they have taken in a day. They also track calories burned and quality of sleep. They even monitor heart rate. Activity trackers inform people if they have fallen behind their exercise goals. They motivate them to become active if they are spending too much time sitting around. Other companies have noticed how important these devices are: in 2019, Google bought Fitbit, whose IoT bands are worn by 28 million people, for approximately $2 billion.

There are even smart **compression** leggings and shorts that can detect several things. They can tell how hard your muscles are working or if one leg is stronger than the other. They can also tell if the correct muscles are being engaged based on the workout.

Those looking to maintain better eating habits and increase their exercise may benefit from smart belts. The Welt belt does more than count steps. It detects tension through a sensor in the buckle, which tracks overeating. Users can also track whether they've gone up or down a waist size over time.

Those who wish to keep eating might try the Belty. It loosens automatically in cases of overeating. It adjusts itself based on weight gain or loss. But it does not allow poor health habits to go undetected. It gives feedback on how to lead a healthier lifestyle.

The SMART (Seizure Monitoring and Response Transducer) belt has a more specific function. It provides greater peace of mind for those suffering from **epilepsy**. It is fitted with two **electrodes** to detect sweat on skin and changes in breathing. It then wirelessly sends messages to friends or family members to respond.

IoT technology has impacted what is worn on your feet. Sensors in the soles of shoes analyze stride and posture. The

Some smart headphones can translate languages for you!

smart shoes then advise users how to reduce fatigue and chance of injury when walking or running. Smart shoes also detect if a person is moving on flat ground or on an incline. This allows the device to count calories burned. They can even sense when the laces are loose! Users can tighten them with a click on the smart shoes **app**.

IoT wearables can make a positive impact beyond health monitoring. There are wearables that automatically adjust based on the user. For instance, Nike has shoes that tighten or loosen themselves based on the user's feet. This is great for athletes whose feet may swell during an intense game or workout. There are even IoT-connected wearable devices, like Embr Wave, that act as personal **thermostats** to warm or cool you down!

Some smart wearables are also fashionable. For instance, there's a smart ring that not only acts as a fitness tracker, but also alerts users that their **Uber** has arrived. It even has a built-in panic button in case of emergency. Smart apparel is being produced by apparel companies like Nike, Under Armour, and Ralph Lauren. The advantage of this apparel is that it tracks biometric data and adjusts to suit your level of comfort or exertion. For example, PoloTech shirts at Ralph Lauren can measure your heart rate, breathing patterns, energy levels, and more.

The gaming industry also benefits from smart wearable devices. Video games no longer have players simply pushing buttons in front of a TV. Players now have the option to wear headsets and take the game to the great outdoors (or to their backyard) and use their body movements to react to the game. There are even specially made clothes that make the experience of video games that much more realistic!

IoT can also impact education. For instance, smart gloves connected to an IoT system help teach sign language to children remotely. Headsets and watches report brain activity of students to a mobile app. This allows teachers to monitor student focus and understand how to get through to them.

Smart wearable devices bring ease to daily living. But are there drawbacks that outweigh the benefits?

NO ORDINARY GLASSES

Eyeglasses are not just for improved vision anymore. Glasses linked to the internet have many uses. For instance, smart glasses have a camera built into the lens. This allows users to take photos or videos and send them wirelessly to a smartphone or laptop. Some camera functions are controlled with the click of a button. Others need only a verbal command. Smart glasses also blend augmented reality (AR), which can help in navigating to a new area.

However, there are still a few issues to address. For instance, people are worried that smart glass users can take their picture without their knowledge. Another issue is cost. Google stopped selling its Google Glass model that cost $1,500. The much cheaper Snap Spectacles, selling at $380, seemed too expensive as well.

Smart glasses have the potential to provide far more important uses. They might allow people with visual impairment to gain back their sight. They could aid the hearing impaired. It is believed they could even detect cancer cells inside the brain.

Some smart glasses incorporate augmented reality (AR) technology into the glasses.

CHAPTER TWO

Drawbacks to Smart Wear

Revolutions don't happen overnight. This is certainly true with IoT and smart wearable devices. Problems range from cost to functionality to looks. Potential users of wearable devices weigh the cost against thc benefits. People may want to wear a ring that alerts them when their rideshare has arrived or play video games using a sensor-controlled vest and headset. But they might not like the price tag. The previously mentioned PoloTech apparel from Ralph Lauren costs almost $300.

Early market trends show that people mainly want wearables to promote a healthy lifestyle. Activity trackers help ensure that they are getting a needed amount of exercise. They monitor heart rate and breathing patterns. They can improve how people walk and run. While most wearables are health- and fitness-focused, there is still potential for other industries, like gaming, healthcare, and education. But if the wearables that aren't focused on health or fitness aren't selling, the price tag might

Virgin Atlantic, a British airline, improved customer service by using a smart wearable device to help greet and provide personal care to its first-class passengers.

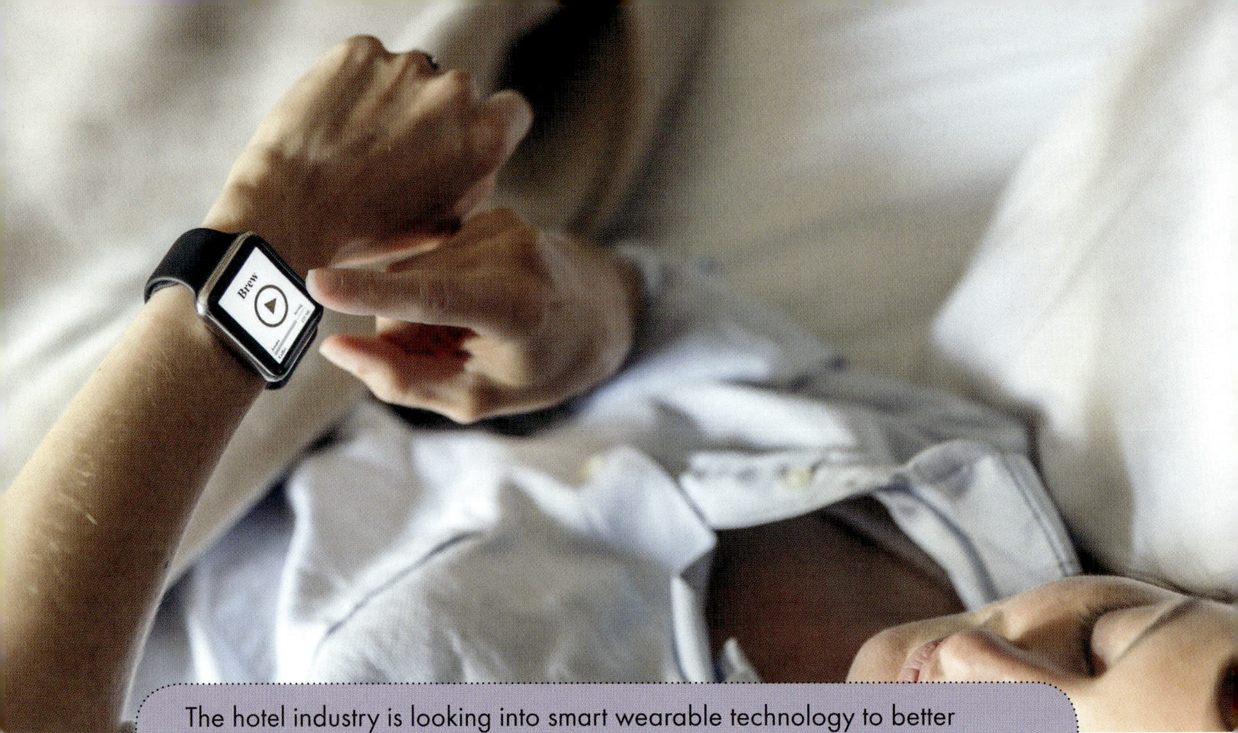

The hotel industry is looking into smart wearable technology to better serve guests.

remain out of reach. One hope is that research into wearables will focus on other industries beyond gaming, healthcare, and education.

There are other practical issues. One revolves around compatibility. Devices and sensors made by different **manufacturers** might not be compatible with each other. For example, controlling a home appliance from a wearable device requires a network connection and compatible software. Another issue is privacy and security concerns. Smartwatches and other smart wearable devices collect a good amount of personal data. This data—like location, health, and financial information if the device is payment-enabled—can be potentially hacked, sold, and **exploited** for a profit. In 2019, the

Royole Corporation, a technology company, introduced the first smart clothing to include flexible high-definition displays.

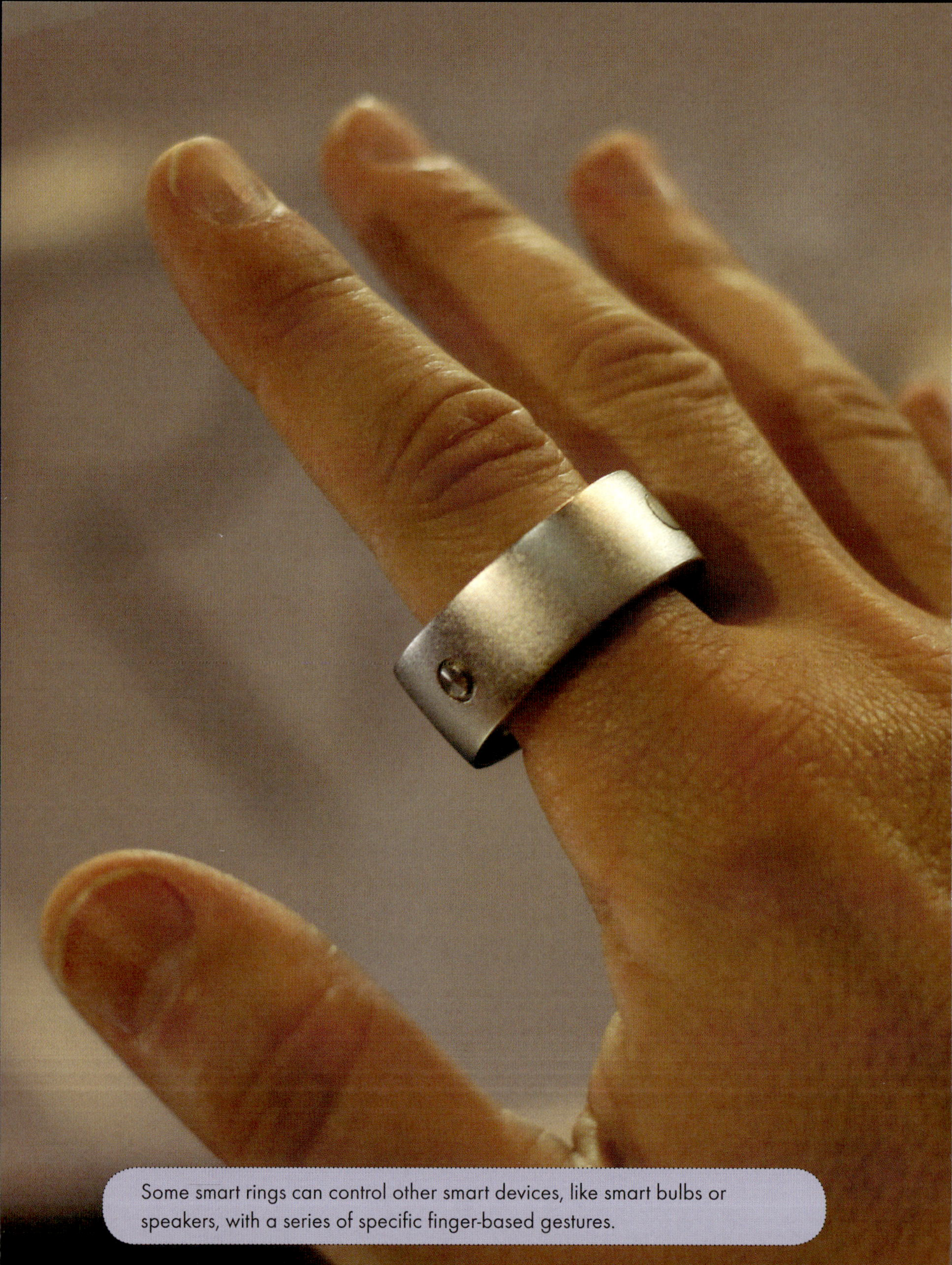

Some smart rings can control other smart devices, like smart bulbs or speakers, with a series of specific finger-based gestures.

U.S. government's cybersecurity budget was $15 billion. Consumers need to remember that all smart wearable devices store their data on the cloud. So, a potential data breach is always a possibility.

There are rings and then there are smart rings. These rings have been created with sensors. Some people like them because they are lighter than smartwatches and bracelets. And they look like jewelry. The smart rings connect to smartphones through Bluetooth. That allows users to learn more about the data they send.

Many smart rings track health. They monitor pulse and heart rate. They can also track sleep and activity. Other smart rings help people communicate. They notify users of text messages or missed phone calls. They display social media alerts from a smartphone.

THE BIG CHALLENGE

Tech giant Intel had a big idea in 2014. It offered $500,000 to the best plan for a new wearable device. The company eventually narrowed the entries down to 10 finalists. The grand prize winner invented a unique wearable called Nixie. It was a small drone worn like a wristwatch that took photos and videos.

Nixie had four flexible bands that were released with a sweeping arm motion. It caused a drone to fly off the wrist into the air. The drone would then take a picture or video and return to the user like a boomerang. It could be programmed to hover or even follow a moving owner. Once Nixie returned to the wrist, its sensor would connect to the user's smartphone. The photo or video it took in flight would immediately upload to the mobile device. That made Nixie useful during activities such as biking and rock climbing.

The Nixie drone could indeed take off. But the wearable did not take off as a commercial success.

There are even smart devices for baby diapers that can detect and alert a caregiver when a diaper needs changing.

CHAPTER THREE

The Future of Wearable Devices

Two decades into the 2000s, the popularity of wearables has extended little beyond activity tracking. But that doesn't mean research and **beta** products aren't underway in other industries.

Consumers could embrace IoT technology in gaming. Manufacturers are sure to offer many games that require wearables both inside and outside the home. And companies are certain to come up with ways to make everyday chores as easy as pushing a button on a smartwatch from miles away. For example, according to tech company iRobot, in 2016 as many as 20 percent of the world's vacuum cleaners were robotic.

But wearables can make the biggest impact in the fields of education and healthcare. There are schools that are looking to invest in IoT wearable technology to track students' engagement and learning styles. The data collected by the smart wearable device would give teachers feedback and suggestions on how best to serve the student. Some schools are also incorporating

There are smart wearable devices, like headbands and headphones, that are designed for better sleep.

Experts suggest that wearable technology can help students focus and that it can foster innovation and creativity.

IoT technology into the student ID cards. (Technically, this isn't a wearable device, but students generally always have their IDs on them.) Some IoT-connected IDs have sensors that let parents know where their students are during emergencies. Other smart student IDs track attendance.

Doctors and patients also benefit from smart technology. For instance, researchers are currently working on smart contact lenses that can monitor blood sugar levels. This could help people living with diabetes. The potential of smart lenses doesn't end there! Smart lenses could offer more personalized treatments by giving doctors even more accurate and timely data on things like eye fluid, which can reveal whether you have certain diseases like glaucoma or diabetes. These lenses could

Mojo Vision, a startup in Silicon Valley, California, is developing a smart lens that might help people see in the dark!

Sure Petcare developed a device that acts like a FitBit for dogs. The device monitors things like exercise, sleep patterns, and even how much your dog eats.

also allow patients to take **preventive** action by monitoring their health. Such devices could save millions of lives.

One can make predictions on how wearable devices will impact the future world, but only time will tell. The potential of IoT wearables has few limits.

Humans are not the only creatures that can benefit from wearable IoT devices. Wearables monitor the health of pets and livestock. These devices can detect heart rate and body temperature. They can alert owners if a pet is getting sick. They can count calories burned on walks. They can transmit data to farmers about grazing patterns to keep cows healthy.

Wearables such as collars and tags can also ensure that animals do not stray. They can be fitted with **GPS** trackers. Owners are told on smartphones or home computers when an animal has wandered off.

GOOD VIBRATIONS FOR THE DEAF AND BLIND

It is hoped that the Internet of Things can bring a new world to the deaf and blind community. Devices have been developed that could improve their senses.

One wrist wearable could transmit patterns of vibrations. Those vibrations would allow deaf people to understand speech. They would associate the patterns with sounds.

Another project revolves around a wearable shirt. The clothing uses sensors to translate music into vibrations over the surface of skin. The deaf would then enjoy a concert experience.

Wearables that transmit vibrations promise to help the blind as well. One such device is the BuzzClip. It is worn on an article of clothing. It uses **ultrasound** to detect obstacles as a person walks inside or outside. It vibrates when a blind person approaches that obstacle.

Bose, an audio equipment company, created smart frames that benefit the blind community. The device uses sound and GPS to help people hear what they're "seeing." Some call it "audio augmented reality."

Critical Thinking and Problem Solving

One problem for the IoT wearable industry is convincing people they need these products. Many consumers believe such devices are too expensive. Some feel there is no big need for wearables in their lives. Others do not like the way they look.

Companies that produce wearables will spend billions of dollars in the future on advertising campaigns. They will try to create a market for ordinary people. But it will not be an easy task. New technology takes time to be accepted.

Think About It

Do you think IoT wearable devices are worth the price? Why or why not? In what industry do you think these devices will have the most impact?

Amazon, a technology company, is developing a smart wearable that can detect how you feel from the sound of your voice.

Learn More

Books

Cohen, Sahrye, and Hal Rodriguez. *Make It, Wear It: Wearable Electronics for Makers, Crafters, and Cosplayers*. New York, NY: McGraw-Hill, 2018.

Gitlin, Martin. *Wearable Electronics*. Ann Arbor, MI: Cherry Lake Publishing, 2018.

Websites

Wonderopolis—What Is the Internet of Things?
https://www.wonderopolis.org/wonder/what-is-the-internet-of-things
Discover more about what IoT is and its devices.

Wonderopolis—What Kinds of Technology Do Athletes Wear?
https://www.wonderopolis.org/wonder/what-kinds-of-technology-do-athletes-wear
Learn more about the smart wearable devices that athletes wear.

Glossary

app (AP) a computer application for mobile systems and devices

beta (BAY-tuh) the version of a product that is almost finished and is used for testing

cloud (KLOUD) data that is stored on the internet

compression (kuhm-PRESH-uhn) clothes that fit tightly against the skin to help blood circulation

data (DAY-tuh) facts to be used for planning or making decisions

electrodes (ih-LEK-trohdz) parts of a device such as a battery that allow electronic currents to come and go

epilepsy (EP-uh-lep-see) nervous system disorder that causes convulsions and an unconscious state

exploited (ek-SPLOIT-id) used something for one's own advantage

GPS (JEE PEE ES) a system that uses satellite signals to determine a user's location and give directions; stands for global positioning system

manufacturers (man-yuh-FAK-chur-urz) companies that produce things in large quantities

monitor (MAH-nih-tur) to check on or keep track of something

preventive (prih-VEN-tiv) measures taken to prevent a negative outcome

sensors (SEN-surz) detection devices that respond by transmitting a signal

technology (tek-NAH-luh-jee) related to applying knowledge about science and industry

thermostats (THUR-muh-stats) devices that control indoor temperature

Uber (OO-bur) a rideshare company that transports passengers

ultrasound (UHL-truh-sound) ultrasonic waves for a medical diagnosis

Index

activity trackers, 6, 12, 20
Amazon, 29
animals, 24, 25
audio augmented reality, 27
augmented reality (AR), 11

biometric data, 9
blind community, 26, 27
blood sugar, 22
Bose, 27

cameras, 10
clothes, smart, 9, 15
compatibility, 14
connection, 4–5
contact lenses, 22–23
customer service, 13, 14
cybersecurity, 17

data, 5, 9, 14, 17, 20, 22
deaf community, 26
diabetes, 22
diapers, smart, 19
drone, 18

eating habits, 7
education, 5, 9, 12, 20, 22
Embr Wave, 8
epilepsy, 7
exercise, 6, 7, 12

farming, 6
fashion, 9
feet, 7–8
Fitbit, 6
frames, smart, 27

gaming, 5, 9, 12, 20
glasses, smart, 10, 11, 27
gloves, smart, 9
Google, 6, 10

headphones, smart, 8, 21
headsets, 9, 12
health, 5, 6, 12
health trackers, 17
healthcare industry, 6, 12, 20, 22–23, 25
hotel industry, 14

Intel, 18

leggings, compression, 7

Mojo Vision, 23
muscles, 7

Nike, 8, 9
Nixie, 18

pets, 24, 25
PoloTech shirts, 9, 12
privacy, 14

Ralph Lauren, 9, 12
rings, smart, 9, 12, 16, 17
robots, 20
Royole Corporation, 15

security concerns, 14, 17
senses, 26
sensors, 4, 7, 14, 17, 22, 26
shirt, smart, 26
shoes, smart, 7–8
sign language, 9
sleep, 21
SMART (Seizure Monitoring and Response Transducer) belt, 7
smart wearables, 4–5
  benefits of, 6–11
  costs, 12, 14
  drawbacks, 12–18
  future of, 20–27
  practical issues, 14
  what people want, 12
smartwatches, 6, 14

travel, 6

ultrasound, 26
Under Armour, 9

video games, 9, 12
visual impairments, 10
voice, 29